The Watery Places of Suffolk

Pip Wright

Other books by Pip Wright

Lydia
Death Recorded
I Read it in the local Rag (pub. by Poppyland Publishing)
Exploring Suffolk by Bus Pass
Thomas Slapp's Booke of Physicke
A Picture History of Margaret Catchpole

Books by Pip & Joy Wright

The Amazing Story of John Heigham Steggall,
'The Suffolk Gipsy' (ed. by Richard Cobbold)
Newspapers in Suffolk (6 vols)
Grave Reports
Witches in and around Suffolk
Bygone Cotton

See all these at **www.pipwright.com**

&

The Diary of a Poor Suffolk Woodman
(with Léonie Robinson, pub. by Poppyland Publishing)
see **www.poppyland.co.uk**

ISBN 978-0-9564855-0-2
Published by **Pawprint Publishing**
14, Polstead Close, Stowmarket, Suffolk IP14 2PJ

The Watery Places of Suffolk

My wife believes that I must have been a labrador in some past life. Wherever I might be, I gravitate towards water. Ponds, lakes, rivers, marshes or the sea - these are the places I love. I firmly believe that a landscape is incomplete without water forming a key part of it.

Assuming I am not the only one who thinks this way, I dedicate this book to all seekers of wet places... to those who love to meander through marshes, to flounder in fens, to ramble by rivers, languish by lakes or simply saunter beside the sea; this book is for you.

The Waveney near Homersfield

Where there is water, there is often wildlife in abundance. It is not by coincidence that a large number of Suffolk's nature reserves sit alongside and within the damper parts of the county, beside river or coast; amidst fen or reedbed. And when you chance that way, who knows what you might be fortunate enough to find.

Water Rail at Fisher Row, Oulton

So here are some of my favourite places. Some were last visited in times of summer drought; others when the water was hard frozen and the ducks struggling to find anywhere to swim. Believe it or not, there are still places described here that, whatever time of year you might go, with any luck you'll be the only one there.

Most of the photographs are my own, but I am indebted to wildlife photographer Richard Weale for the use of eight superb pictures that have added so much to this book.

All of the maps referred to in this book are from the Ordnance Survey Explorer series.

Suffolk river haunts

Suffolk is a county defined by rivers. Indeed, most of the county boundaries follow rivers - not great nationally-known rivers: altogether smaller, gentler, more tranquil rivers that make their way through woods and water meadows and fen and farm on their way to the North Sea.

For all that, the varieties of riverside vistas available to us are endless. These, though far from all-embracing, are some of my favourites.

To the north of the county near Thetford is an area that is described on maps (O.S. Explorer map 230) as Knettishall Heath. When my children were small, they were delighted to discover a place where river swimming was (and still is) very much possible. This is the Little Ouse river, on its way to join its larger namesake. The water is cold and clean and lovely, and the Breckland wildlife stunning.

Santon Downham (pictured below) to the west of Thetford is another place where we as a family have enjoyed many days bathing, canoeing and just ambling along the river banks.

Travel a little farther west and you come to the town of Brandon where it is possible to have coffee and cake by the riverside and hire a boat to row gently down along the Little Ouse, much of the way through woodland as you head for Santon Downham. (O.S. map 229)

Still in Suffolk (just) and still in the vicinity of the Little Ouse, to the west of Lakenheath you can find a true wetland wilderness (O.S. map 228). Parking just about a mile along the road towards Sedge Fen, you can pick up a part of the Hereward Way. You walk about a mile and a half along what the map describes as the Stallode Bank, crossing the railway (carefully) and making for the river. First, you encounter a vast reedbed, the haunt of reed-warblers and sedge-warblers in summer. Next, you pass a graveyard of trees set in a watery mere. This is referred to on the map as Botany Bay and feels as remote as the original must have seemed once upon a time. The dead tree stumps proved perfect perching posts for cormorants when I was there one crisp cold bright January day.

Botany Bay, Lakenheath

Turning right along the river you find you have river on one side and what is called Norfolk Fen on the other. This is a remarkable area for wildlife and you are right in the middle of it. It must be one of the wettest places on earth. Everywhere is moist - and full of wildlife. And I found a whole bank of snowdrops beside a frozen pool, struggling to come into bloom, just feet from the river's edge. You notice just how high the river is above the surrounding countryside. That is the Fens for you! A small pumping station works night and day raising water about ten feet from the dyke below to the river above. From there, you can continue along the river, but it is quite a trek. Eventually, you can make for the Suffolk Wildlife Trust Field Centre at Lakenheath.

The Little Ouse rises at much the same point as the Waveney which flows in the opposite direction. This is a marshy area known as Redgrave & Lopham Fen (see page 51). The Waveney valley separating much of Suffolk from Norfolk is a truly lovely area.

Redgrave Fen

If you visit Bressingham Gardens in Norfolk, a couple of the railway excursions take you down beside the Waveney where it is little more than a stream. Further down, past Diss, the river meanders through wide, lush water-meadows. Between Hoxne and Weybread are some lovely stretches of river. Several small roads and tracks cross the Waveney into Norfolk at points where old water-wheels once turned. It is possible to park just north of the river beween Syleham and Brockdish (O.S. map 230) and walk two miles of the Angles Way to the Mill Road at Weybread. You are never very far from the river, and I have seen barn owls and herons amongst the sights worth noting.

As the Waveney moves and grows on its way to Bungay and Beccles, you can find paths that offer spectacular views. Part of the Angles Way between Mendham and Homersfield follows higher ground, just above the flood plain, and is not to be missed (pictured on page 3).

Bungay and Beccles are just a few miles of the Waveney apart, but they could not be more different. Beccles shouts about its river. It is as if it exists simply because the River Waveney is there. You can take trips, hire boats; even live on the river. You can access the Broads from there. Or you may simply choose to amble along the riverbank all the way to the sea. (see right)

At Bungay, (left) it is as though the river is a well-kept secret. It seems hidden away from the rest of the town, not quite so obvious and rather less navigable, but it is there nevertheless. You can walk its banks, with the town's church towers rising up behind as you do so.

Better still, take a circular stroll along what O.S. map OL40 describes as 'Bath Hills Road.' (to the north of the town, off the Denton Road). This takes you round one great loop of the river, bringing you past lakes, through woodland and round to Ditchingham before returning to the town. The whole walk is a good three miles, and worth every step.

On the banks of the Waveney

There are tracks at North Cove, Carlton Colville and Oulton whereby you can reach the river. I'm always drawn to roads with titles like 'Marsh Lane' or 'Fen Street'. You'll find a few of those round here.

At Marsh Lane, Worlingham, you can drive down as far as the railway crossing. There may be room to park. From there it is half a mile or so to the river. A towpath-like trail runs along the Suffolk bank of the river as far as Oulton Broad. The land beside the river is low marsh and water meadow intersected by dykes. Down at Carlton Colville is a splendid marshland nature reserve.

Herringfleet and St. Olaves (in Norfolk) to the north east of the county (O.S. map OL40) have a lot to appeal to the water seeker. You'll find a restaurant in the grounds of the old priory and the Bell Inn is close by the River. Boat hire is available - it is about 2 miles downstream to the Dukes Head at Somerleyton. The railway there crosses the river by way of a precarious-looking swing bridge. The footpaths that run through Somerleyton Marina and Boatyard give you great views of the bridge and this section of the river.

But if you are full of energy and keen to see much of the watery land between Norfolk and Suffolk from the seat of a bicycle, part of the Regional Cycle Route 30 (also known as the Two Rivers Route) is for you. The section from Lowestoft to Diss deviates between the two counties as it follows (fairly closely) the River Waveney. After Diss, it follows the Little Ouse valley along country roads to Thetford and Brandon. You can download maps of the route by visiting **www.sustrans.org**.

Now we plunge south to Suffolk's other great river border. Dividing most of Suffolk and Essex, the River Stour has the power to captivate, wherever along its length you might find yourself.

The source of the river is probably just to the north of Great Bradley, but by Kedington it is clearly recognisable as a river. At each of the river crossings, you tend to find a rather attractive village, such as Wixoe and Stoke-by-Clare, nearly all with an old mill building.

In the little town of Clare (O.S. map 210) you have so many good things in a small area - the old railway buildings sit enticingly between the river and castle mound. There is a priory and antique centre, all within a stone's throw of one another, and the town of Clare full of tea-rooms and pubs nearby. The Tourist Information Centre in the old railway station will furnish you with a map illustrating river walks along this stretch of the Stour.

Clare castle

The middle reaches of the River Stour are lovely. A splendid section of the Stour Valley Path passes west of Long Melford, leading down to Rodbridge. Here, there are pools beside the river teeming with wildlife. But evidence of just how wet this area can get is shown in this raised path leading across the river to the Essex hamlet of Liston. It is designed to keep your feet dry in all but the very worst of flood conditions.

Check out the water meadows at Sudbury (unfortunately boat-hire at Ballingdon Bridge is temporarily suspended).

The Stour Valley paths from Bures to Wormingford or from Wissington to Nayland are also to be recommended (O.S. map 196). Park near Wissington (pronounced, and sometimes spelt 'Wiston') Church. (shown here)

This is probably the quietest place I know. It is altogether enchanting. It's the kind of place I was reluctant to mention, for fear too many people might discover it.

The twin villages of Bures and Nayland derive their attraction from their closeness to the river. At Nayland, a number of the houses back on to the Stour and it is almost obligatory to have a small boat at the end of your garden.

The River Stour from Bures Bridge

The Stour is just as interesting when it becomes a tidal river. Starting from Cattawade on the Suffolk side of the Stour from Manningtree, you can walk the tidal reaches or explore the paths upriver to Flatford Mill (O.S. maps 196 & 197). A National Trust tea-room awaits you. The Dedham vale area is well known and you'll find on a summer's weekend you are walking the area in company with a host of other people.

If you prefer to explore other tidal reaches of the Stour, park at Lower Holbrook and make for the river. Alternatively, using as your base the Bakers Arms at Harkstead on the Shotley peninsular (O.S. map 197), you can use footpaths to reach creeks in Holbrook Bay on the widest part of the river. Continue to Stutton and you will reach the Kings Head. Both pubs sit beside bus routes that link with Ipswich.

From a footpath beside Erwarton church, you can walk around the creeks to Shotley where the rivers Orwell and Stour meet, the docks of Harwich and Felixstowe clearly visible across short stretches of estuary. (Pictured right: Holbrook)

But, of course there are all the other lovely rivers of Suffolk, contained well within its borders, that are not to be ignored.

If a riverside walk is your idea of heaven, start from the edge of Cavenham Heath Nature Reserve (O.S. map 226) by Temple Bridge and walk part of the Lark Valley Path towards Mildenhall. Expect to see a host of birds, wildflowers and water mammals. If you are driving to there, you can park close to this point by approaching along lanes from either Icklingham or from Tuddenham. In Mildenhall itself, (pictured left) riverside walks have been opened up

Suffolk has a host of these riverside trails. The Gipping Valley Path takes you from Stowmarket to Ipswich, but the best part has to be from Needham

Lake to Great Blakenham (O.S. map 211) where you pass any number of lakes that have been produced by gravel extraction. This stretch is a haven for fishermen, and for waterfowl like this mute swan (pictured right).

From the point of view of wildlife watching, try the Alde river path from Snape to Iken and circle back by way of the Suffolk Coast Path. Depending on the season of the year, it is likely you will see wading birds, water-fowl and any number of smaller birds that flit in and out of the reeds.

Iken church beside the Alde

Remember however, with all close-to-river tracks, there will be times when these may prove impassable. Near to the estuary, tides may come into play, and just because tracks are shown as footpaths on a map does not guarantee their existence at all times of the tide.

Sometimes the wildlife along a river catches you by surprise, as can be shown with this picture I took of a fox sunning himself beside the River Dove near Eye.

Here, you can park near Abbey Bridge and walk along the marked footpath (O.S. map 230) upriver past the old priory. Lost amongst the trees are the old monastic fish-ponds. Soon you find yourself on the Bolser Bridge amidst water meadows. A short walk enables you to circle round to the little town of Eye with its medieval castle mound and lovely church... a short watery walk with a thousand years of history attached.

If you prefer a little more company and enjoy your rivers from a town perspective, visit Ipswich Docks. This is a rapidly developing area and changes appear every time you visit. Here historic architecture sits comfortably side by side with the new.

You don't even have to be particularly energetic to enjoy a view of the river. Sit outside Isaac's with a pint in hand and take in all there is to see. Trips up the Orwell can be booked aboard the Orwell Lady (see **www.orwellrivercruises.co.uk**) at Ipswich Tourist Information Office. The Orwell from Ipswich to the sea is spectacular as you sail under the Orwell Bridge towards Felixstowe.

Several of Suffolk's rivers can be enjoyed from on board a boat. The Lady Florence, described on page 30 gives you a great view of the lower Alde. Higher upstream, short cruises run during the summer months from Snape, beside the Maltings.

Trips depart from Waldringfield Boatyard, cruising the Deben (**www.ukattraction.com/east-of-england/deben-cruises.htm**).

Topsail Charters (**www.top-sail.co.uk/about_us.htm**) run a programme of sailing barge trips exploring the Orwell & Stour rivers.

The Waveney Stardust is a Broads motor cruiser specially designed to enable the disabled to enjoy a river trip. Full wheelchair access is available. A number of their cruises depart from Oulton Broad (**www.waveneystardust.co.uk**). Further upriver, the Liana runs boat trips from the quay at Beccles (**www.broads-authority.gov.uk**).

One Orford boat not available for excursions

Occasionally the steamship, the Balmoral runs sea-going excursions and the last sea-going paddle-steamer in the world, the Waverley runs trips from Ipswich to London and the Thames (**www.waverleyexcursions.co.uk**). This is a trip well worth looking out for as it offers stunning river, estuary and sea views.

Places offering food and refreshment beside Suffolk rivers include the Ramsholt Arms on the side of the Deben, the Butt & Oyster at Pin Mill on the Orwell and the Maybush at Waldringfield where the views are quite magnificent. Then there is the Locks Inn at Geldeston near Beccles. This is technically in Norfolk, but sits beside the Waveney and is easily reached in Summer by ambling the two miles along the Norfolk side of the river or by footpath from Shipmeadow.

The Ramsholt Arms

Tidal stretches mean either the view may contain a lot of water or, alternatively, a lot of mud. Often mud is best if it is wading birds you are after. Blythburgh can be a haven for oyster-catchers, redshank, curlew, and shelduck when the tide is out. For the energetic among you, walking from there to Walberswick to the south of the River Blyth offers magnificent views and plenty of wildlife to savour (O.S. map 231).

Looking across the Blyth near Walberswick

At Walberswick, you can cross the river by ferry or by using the old bailey bridge. Once a railway ran across the river by way of a swing bridge, but that closed as long ago as 1929. From there you can make your way into Southwold where good bus services operate.

When it is dry enough, you can walk inland along the Blyth from Blythburgh to Blyford Bridge (wear strong waterproof footwear!). This is only a couple of miles and brings you out near a pub (Queens Head) and on a bus route (Service 520)

I like the idea of an oasis at the start or end of a river walk. The Ship at Levington (O.S. map 197) is situated above the Orwell foreshore. Footpaths opposite lead you down to the shore and either west to Nacton or east to the wonderful Trimley Marshes. At Ramsholt, on the side of the Deben, food and drinks are available from the Ramsholt Arms. Walks from here along the river foreshore can bring you to one of the best fossil hunting areas of Suffolk.
(see: **www.ramsholt.ukfossils.co.uk**)

At Woodbridge (O.S. map 212), you can follow the riverside for the length of the town, taking in Marina and Tide-Mill. You'll find an excellent restaurant beside the latter. Then, follow the river to Wilford Bridge where the pub serves excellent food. The other side of the bridge offers a wildlife trail to die for. Redshanks and sandpipers paddle in the mud, kingfishers and reed-warblers are plentiful here. Don't forget your binoculars.

Woodbridge Tide-Mill

And if all that sounds a little over-exuberant, and you yearn for a flat, undemanding, but thoroughly lovely walk, follow the River Brett at Hadleigh from the Council Offices to Toppesfield Bridge and back (O.S. map 196).

Or just let the beauty of these places take your breath away. Find the church on the hill at Hemley near Newbourne (O.S. map 197) and stand and look out across the Deben...

All Saints church,Hemley

....Or find one particular hidden track at Syleham (O.S. map 230) and take in the beauty of the Waveney valley. Most churches stand up above most of the surrounding countryside. The tiny round-towered church of St. Mary's at Syleham is tucked away from it all among the Waveney water meadows, and you can picture a time when it might have been simpler to attend church by boat than on foot...
...Or stand in the churchyard at the ancient site that is St. Botolph's at Iken (O.S. map 212) and survey the Alde valley. A little further upriver is a large picnic area with the finest views in Suffolk.
...I defy you to find any more idylic spots than these.

Less frequented than the Woodbridge riverside, but equally lovely, parking up by Martlesham Church gives you footpath-access to the creek where the River Fynn joins the Deben (pictured below).

And well upriver, much nearer the source of the Deben, you have the lovely valley area between Cretingham and Brandeston and on down to Easton, a beautiful area intersected by small roads and footpaths.

The water-spash at Dagworth

At Oulton, near Lowestoft you'll find tracks down to the Oulton Marshes where the wetter it becomes, the more the wildlife likes it. I could go on and on.

It's encouraging to know that this county is still full of secret watery corners just awaiting our discovery.

The Lakes of Suffolk

I know, I don't live in the Lake District and it would be easy to assume that there are few large stretches of water in Suffolk to interest people. Think again!

Alton Water (pictured below) is not a natural lake: it is a reservoir. But it is home to wildlife aplenty and it is very accessible. At the Stutton end, you can park, hire bicycles and get a cup of coffee. At the Tattingstone end, you can park and find a pub not far away. What is on offer here is a glorious lakeland country park with an eight mile track around it. Summer or Winter, it is strongly recommended (O.S. map 197).

West Stow near Bury St. Edmunds (OS map 229) is best known for its reconstructed Saxon Village but it also has free parking and access to walks around the country park where you can enjoy river and lakeland scenery. The Lark Valley Path to the east leads past the lakes at Culford School. Across the other side of the river at Lackford is a wildfowl nature reserve where all manner of unusual water birds congregate in the flooded gravel pits.

Tufted ducks at West Stow

It is worth a mention that at West Stow Country Park, you'll find toilets, and a tea-room with a great view of the bird-feeders.

Nearby are lovely walks around Ampton Water and Livermere (O.S. map 229). You can park near Great Livermere church and follow the footpath through two small gatehouses towards the water. A fair bit of the land at this end of the Broad Water is scrub - good for ambling down to the water's edge. This area is full of geese in winter and the air resounds with skylarks in summer. Away at the opposite end of the lake, you can see the tower of the ruined church at Little Livermere. This is one of the lesser known beauty spots of Suffolk and definitely one not to be missed.

Geese at Livermere

Thorpeness (OS map 212) near Aldeburgh boasts a lake known as The Meare. Here you can do the whole 'Swallows and Amazons' thing. You can hire a boat and explore the islands dotted across the lake. There is a tea-room by the lake and a good pub nearby.

Thorpeness

The lakes at Minsmere Nature Reserve to the north of there have been artificially created to attract the best variety of wildlife anywhere in Britain. You will need to obtain a permit to visit, but it is well worth it. From the numerous hides, you'll be astounded by what you can see. You can approach Minsmere from Eastbridge or from the Westleton to Dunwich road. Either way it seems an age before you get there, eventually arriving at the Field Centre, where there is plenty of parking, a shop and cafe... and of

course a wonderful place to watch birds (and not just birds as these pictures show). I last visited Minsmere in January, when you could see pintail, widgeon, shoveller and bearded tit. Two people (not me, unfortunately) recorded sightings of bittern. In Spring and summer, there is even more to see.

More publicly acessible are the lakes at Weybread, south of Harleston in the Waveney Valley (O.S. map 230), where coots and

Weybread

grebes, reed-warblers and herons once shared these pits with sand extractors. Now this is the home of the fisherman, but the wildlife is every bit as plentiful. Here is the real lakeland of Suffolk. The waters stretch away into the distance and it is perfectly lovely here whatever the season.

Fritton Lake, to the north of the county (O.S. map OL40) makes the most of a beautiful setting. As well as the natural beauty of a lake in the woods, you have boat hire, floating guided tours, nature walks and pony rides plus much more. It's a really good family day out. There are places to eat and drink and the scenery is lovely, though it can get crowded at peak times. The main lake is actually just one of a series that crop up in the vicinity and you may encounter them as you cycle or motor around the roads just north of Lound.

If you prefer less of a tourist trap, go in search of Framlingham mere (O.S. map 212). It sits below the castle on one side and is overlooked by Framlingham College on the other. A footpath encircles the mere, but at some distance from it. This small nature reserve is managed by the Suffolk Wildlife Trust on behalf of the College.

Cornard Mere - with reed bunting inset

Something of a surprise is Cornard mere: (O.S. map 196) close to a heavily built-up area, it is a haven of tranquility just off the busy Sudbury - Bures road. Reed buntings, grass snakes and dragonflies are all abundant here. As this picture shows, you get a great view from the hill above.

There is a growing number of ponds and lakes created for the fisherman. Several of these offer refreshment for those who fancy a break from casting lines or those who just like to sit and watch them do it. Many of these fishing ponds and lakes are home to a plethera of wildlife. At Hinderclay, Walpole, Kirton and Stowmarket you'll find such lakes. At Lakeside, close to Onehouse Shepherd & Dog, you can get a good breakfast whilst looking out on the lake. The view at the cafe beside the Suffolk Water Park at Bramford is even better. Coots and grebes dive for their breakfast, as you have yours served to you at a table whilst watching them.

Hinderclay

Through the years stately homes and some of our greater houses have seen landscaping on a grand scale, whereby breathtaking vistas have been created. Many of these are tucked away on private land, inaccessible to the likes of you and me. However, the 'Invitation to view' scheme has made it possible to visit and enjoy such delights as Crows Hall and Euston Hall. (see **www.invitationtoview.co.uk**) The National Garden Scheme, whereby lovely gardens are opened to the public for charity can enable you to get to spectacularly lovely watery places (see **www.ngs.org.uk**). Under this scheme, I recently visited Barton Mere, near Bury St. Edmunds where the lakeside scenery is stunning.

Barton Mere

All too often, the best views are restricted to those lucky enough to live in lovely places, but fortunately a footpath enables you to walk right through the grounds of Heveningham Hall near Halesworth.

The recently restored lakes there are are a must for all lovers of watery places.

Heveningham Hall

And just by way of something entirely different - why not make your way one Sunday morning to Needham Market Lake, where the local model boat enthusiasts can be seen frightening the ducks and annoying the fishermen when putting their splendid mechanical creations through their paces. (pictured right)

Beside the Suffolk Seaside

I am going to leave this section shorter than it really deserves to be as there are plenty of guide books detailing the attractions of our major seaside towns: Felixstowe, Southwold and Lowestoft.

The North Sea forms a semi-solid border to what is Suffolk and it is altering the shape of the county all the time. You can see almost from month to month the changes caused by erosion north of Southwold, and the loss of much of Dunwich and Aldeburgh are legendary.

By way of compensation, you can go to Shingle Street (OS map 197) and see how the opposite applies. Go there and have a good look at the reality that is Orford Ness spit and see how it differs from what the map shows and how it has grown each time you go.

Shingle Street

So, let's begin there. It is a wild and lovely spot. Shingle-nesting birds such as ringed plover and oyster catcher can still be found breeding there in early summer. You can walk in a southerly direction, past the Martello Towers to Bawdsey. You can carry on past Bawdsey

Manor, home of Radar, to the mouth of the Deben where, subject to season and weather conditions, a ferry will carry you across to Felixstowe. There, you'll find an excellent fish & chip cafe and two fine pubs.

Looking across from Bawdsey to Felixstowe Ferry

Alternatively, you could walk north to Orford (O.S. maps 197 & 212). It is quite a trek, in the shadow of the Hollesley Bay Offenders Institution, beside creeks and along the River Ore. At the Butley River, you'll need the ferry (phone in advance: 07913 672499), before progressing past the castle and into the small town of Orford.

Here are ferries to Orford Ness and rather special all-year-round boat-trips. If you book in advance with the Lady Florence (07831 698298) you can enjoy brunch, lunch or dinner as part of an expedition up and down the River Alde. I've been aboard three times and loved it every time. By the way, bus route 71 links Ipswich with Orford.

An entirely different kind of boating experience comes by way of the Coastal Voyager, a nine-metre inflatable offering exhilarating trips around Sole Bay or longer trips to navigate the Alde to Orford or see the seals at Scroby Sands (**www.coastalvoyager.co.uk**).

I've lost count of the number of times I've walked from Dunwich to Walberswick or vice-versa. You can follow the coast along the beach or on the firmer side of the shingle bank. You can instead follow the paths above the Dingle Marshes (O.S. map 231). Either way, you're in for a treat. Marsh harriers and avocets, bitterns and otters make their homes here. Should you finish at Dunwich, I recommend the garden-centre tea-room for such delights as real home-made soup and cake.

Covehithe is a must for the water lover (O.S. map 231) It is remarkable how much nearer to the sea the village is than when I first went there. You'll encounter a road that, rather disturbingly, just ends at the cliff edge (pictured above). Park somewhere near the church (which is well worth a look in its own right) and find the path leading down to the sea and what is called 'Covehithe Broad'. This is a wild and deserted part of the coast. After a storm, it's interesting to walk the beach here and see what has been washed ashore.

Benacre Broad to the north of there is a bit harder to reach but it's well worth the effort. It can be approached on foot from Covehithe, Kessingland or from what remains of Benacre village. Here is a nature reserve complete with bird-hide overlooking the flooded gravel pit that lies just behind the shingle bank and only a short distance from the sea.

Benacre

I last went on a warm but blustery June day by walking from Covehithe Church along the now closed road that once led to the shore. Now it leads to the edge of a sandy cliff. To the north, you follow the path that may eventually disappear altogether. Less than a mile along the cliff, you reach Benacre beach, home to nesting terns and avocets, ringed plovers and oyster-catchers. In winter, a variety of water birds are attracted here. You can see dead tree stumps along the beach that once grew at a distance from the shore. But this shoreline is retreating all the time, as those still living just down the coast at Easton Bavents know only too well.

When looking for seaside places to visit, don't ignore Sizewell (O.S. map 212). Yes, there is the ever-present grimness of the nuclear power-station and you may encounter policemen patrolling the beach with automatic weapons. But parking is cheap, it's a fine deserted beach for most of the year, the all-day breakfast at the beach cafe is as good as you'll find anywhere and you can walk south to Thorpeness or north to Minsmere and Dunwich Heath.

Sizewell beach

From the National Trust Centre at Dunwich Heath, you can park, find toilets and tea-room and embark on a wonderful walk that encircles Minsmere Bird Reserve, leads you beside any amount of watery areas and includes one of my favourite Suffolk pubs, the Eel's Foot at Eastbridge. This circular walk follows heath and dyke, returning along the beach. At times it seems wonderfully remote and is about as good as it gets, to my way of thinking.

And just to prove this book is not only about deserted corners of the county; down at Felixstowe Dock, there is a public viewing area where you can watch the boats go in and out. The East Anglian Daily Times lists their movements. If you want a different angle on it all, in the warmer months, at Shotley, you can catch a foot-ferry that will convey you by way of Harwich to Felixstowe Dock. You pass old light-vessels and get good views of the historic port of Harwich.

Which brings me to Aldeburgh, the gem of the Suffolk coast. You can't do much better than to sit on the terrace of the Brudenell with drink and a good meal and the best view in the world. Then, suitably nourished, you can walk south (O.S. map 212) past what was once the hamlet of Slaughden (now lost to the sea) along part of Orford Ness with the North Sea on one side and the River Alde on the other.

Orford Ness with Aldeburgh to the North

Snipe on the marsh

©Richard Weale Wildlife Photography

The northerly end of Aldeburgh, towards Thorpeness, has a lovely marshy section to the landward side of the road where wading birds and wildfowl gather. Back in the town of Aldeburgh, there always seems to be something interesting going on. And all within sight of a vast expanse of sea. Oh, and don't forget, both here and down at the harbour at Southwold, you can buy the freshest and best fish the North Sea has to offer.

Ponds and Moats, Mills and Bridges

Let's not undervalue the old village pond. They were once plentiful in Suffolk: too many now have been lost along with village greens and wide greenswards.

Where ponds do survive, and especially where they have been given a little tender loving care, they add so much to the ambiance of the place. At Bacton, the pond has been cleaned out and remade the focal point of the village. The Haughley pond spills almost onto the road that passes beside it. Behind is a wooded backdrop concealing the old castle mound. At Lidgate, the pond sits below the church on the hill as if deliberately landscaped. At Tuddenham, the pond is tucked away just off the main road, but looks a treat ringed by clumps of bullrushes. There is Redgrave, where, passing the lovely Redgrave lake to reach the village, the village duckpond is just that - home to dozens of ducks. At Tostock the spring flowers around the pond on the green are truly spectacular.

Tostock with Tuddenham inset

But if you've not yet been there, seek out Wickham Skeith (O.S. map 211), where the green around the mere known locally as 'The Grimmer' just asks to host a picnic. Here, in 1645 the Witchfinder General 'swam' five unfortunate women before bringing them to trial at Bury St. Edmunds. Now it seems altogether more peaceful and more inviting. It's a glorious spot. (pictured below and front cover)

The prize for the best of Suffolk's watery greens has to belong to Mellis (O.S. map 230). Acre upon acre of water-meadow sprinkled with ponds is there to explore. The wild flower list is astonishing. Insects and small mammals abound here, as do the owls that feed on them.

I could write a whole book on the subject of village ponds alone. Doubtless there are others I have missed out that may be at least as deserving as those written about here. All I would say is look out for them and treasure them. They are a dying species.

Kersey

Other watery ways enhance our towns and villages. Streams and rivers trickle through beside roads and sometimes cross in the most scenic way possible. At Debenham, a ford lies just off the main street, and at Little Thurlow there is one at the corner of the small green. In Grundisburgh, you can opt to leave the main road and cross the village by way of a ford.

Kersey (O.S. map 196) is renowned for its 'water-splash.'

There are plenty of surviving fords in the county, but few run right through the centre of a village the way it does there.

Little Thurlow

Walsham-le-Willows, Debenham, Kedington and Peasenhall are graced by streams that trickle through them. Dalham (O.S. map 210) owes much of its charm to the River Kennett that runs close beside the road before continuing on through Moulton. Chelsworth has a desperately narrow bridge spanning the River Brett from which this pretty village is seen at its beautiful best.

The River Stour, bridged at Bures and Nayland (O.S. map 196) can be enjoyed as parts of small circular paths through these villages. They are delightful places to visit. But would they be half as interesting without their river? I doubt it.

There is nothing quite like having a house surrounded by water. Kentwell Hall at Long Melford, Helmingham Hall and Wingfield Castle are particularly grand moated homes, but there are many more, such as Otley Hall and Columbine Hall at Stowupland. Some of these open to the public. Most villages have at least one 'Moat Farm.'

Wingfield Castle

Rumburgh Church is rather unusual. It seems to sit within a complex moat. This betrays its monastic past. The church was once part of a priory and the water that surrounds it may once have served as the monastic fishponds.

Sometimes the bridge over a river is every bit as impressive as the natural landscape. Tucked away, almost forgotten, at Cattawade (O.S. map 197) is this fine bridge.

I've already mentioned Knettishall Heath (OS map 230) as a great place to swim or walk or look for wildlife. But just look at this for a fine piece of Victorian construction. And one village downstream from there is another lovely old bridge at Rushford.

Yoxford, as its name suggests was once a fording point for the River Yox (later to become the Minsmere river). An attractive iron footbridge carries a footpath from the village in front of the lovely Cockfield Hall.

Rodbridge, crossing the Stour near Long Melford gets its name from its bridge (see below). Hadleigh, to the east, has a lovely riverside walk along the River Brett, culminating at a three-arched span.

However, the grandest of them all is to the front of Heveningham Hall. There, a footpath runs right through the grounds and over this recently installed bridge.

The four arched fifteenth century pack horse bridge at Moulton (O.S. map 210) has withstood flood and torrent for hundreds of years. It spans the Kennett amidst a lovely village green. By way of contrast, the Orwell Bridge, just east of Ipswich is quite a sight to behold. You can saunter down under it, along Gainsborough Lane, on the northern side, or just park almost beneath it at Wherstead on the southern side and enjoy the wide variety of water birds that are on show, whatever the season. Here you see oyster-catchers beside the reflection of a bridge support.

Then, of course, we have the magnificent Abbot's Bridge at the corner of the Abbey Gardens in Bury St. Edmunds (O.S. map 211). This possibly dates back to the 12th century and is now a Grade 1 listed building.

Similarly graded, but less easy to locate, is the Iron Bridge at Culford (O.S. map 229). If you make your way towards the church (within the grounds of Culford Independent School) and walk along the footpath through the trees towards the river, you pass along

Abbot's Bridge, Bury

a magnificently landscaped section before coming upon one of the finest bridges I know. From it you get great views in both directions. This piece of waterland is home to a wide range of wildlife. In winter you may see pochard and widgeon; in summer, coot and grebes.

Pochard at Culford

At Hoxne (OS map 230), you can park beside the Community Centre and stand on the bridge that they would have you believe marks the point where St. Edmund was taken by the Danes before his martyrdom in 870 AD. Apparently, a wedding party spotted him hiding under the bridge and betrayed him to his enemies. It is now said that couples riding or walking to their wedding will not cross it, necessitating a diversion, as it is supposed to be unlucky to cross there.

Watermills can still be found in a number of places in Suffolk. A few are open to the public. Pakenham (OS map 211) is a village of two mills. The windmill stands on a hill above the river valley in which you will find a still-working water mill. From April to October, you can view the machinery in action and buy some of the stone-ground flour it produces. There is a tea-room and a recently restored mill-house kitchen. Across from Mill Lane is a patch of land that alternates between being damp water meadow and a sizeable mere. It all depends on the season and the rainfall. I visited it at a wet time, when it was temporary home to a host of ducks and geese and herons.

Heron in flight

Tuddenham Mill near Bury (OS map 226) and Kersey Mill (O.S. map 196) near Hadleigh have become rather attractive rural retreats offering a wide range of possibilities... anything from a cup of coffee to good food and accommodation. Sudbury Mill makes a very elegant hotel, as the photo overleaf shows.

Tuddenham Mill

Sudbury Mill Hotel

Baylham Mill

You can encounter some superb water mills in Suffolk that are not open to the public. However, you can enjoy a good view of them all the same. At Baylham, Sproughton, Wickham Market, Kedington and Letheringham are a few of the old mills worth making a diversion to see. Layham Mill near Hadleigh still has its wheel (pictured above).

Equally fine are others more concealed, such as the old monastic mill at Campsea Ashe Priory. And should you choose to visit the gallery beside Withindale Mill at Long Melford, you may notice the lovely old mill and mill-house set beside the mill pond at the very corner of Suffolk's Stour valley (pictured right).

Then, of course, there is the much vaunted Flatford Mill (OS map 196), courtesy of the paintings of John Constable, now a field centre.

The Tide Mill at Woodbridge (O.S. map 212) is open during the Summer months, as is Alton Mill, transferred to the Rural Life Museum at Stowmarket (O.S. map 211). This would be under water by now if they hadn't moved it. The windpump, sited close to it, (pictured left) is not dissimilar to the one that still remains at St. Olaves close to the Norfolk-Suffolk border (O.S. map OL40).

There are plenty of buildings that exist as a result of the water they survey. Southwold lighthouse (O.S. map 231) is now automatic and therefore unmanned. But it is open to the public at certain times throughout the year. It still serves as a waymark for ships navigating close to the east coast.

The Suffolk-Essex coast is dotted with old Martello Towers, built in anticipation of Napoleon's invasion threat. Most are now unoccupied, though a few have been converted into private dwellings. You see some at Felixstowe Ferry and at Shingle Street. The largest and best is near Aldeburgh at the start of Orford Ness (O.S. map 212).

And if you have a great deal of money to spare, then the ultimate in holiday homes is at Thorpeness (O.S. map 212). Known as 'the House in the Clouds', this old water-tower is available to let: it has 5 bedrooms, 67 steps to the games room at the top and will cost you in excess of £3,000 a week in high summer.

House in the Clouds

The owners of the great houses of Suffolk have often created ornamental water features to enhance their views. The canal at Ickworth House near Bury is one such feature. Farther from the house and buried among trees is the lake referred to as the Fairy Lake (pictured here). National Trust members get to enjoy this walk for free.

But of all the lovely places in this book, I find the salt marshes of the Butley River hard to match. Just on the Orford side of the Froize Inn at Chillesford is a well-marked footpath that leads close to an old duck-decoy, down to an R.S.P.B. managed reserve. This spot is home to curlews and bar-tailed godwit, as well as assorted wildfowl.

Wet Places
- Fens and marshes and Water Meadows

Most of the remaining places on my list are for the naturalist, or for the person that just wants to find an unspoilt corner to wander and enjoy.

Tucked away, not very far from the A140 near Eye (O.S. map 230) is the village of Thrandeston. Just outside the village, is a small marshland nature reserve; really just a couple of undrained meadows ringed by wide ditches, but a great spot for snipe and herons, and tadpoles in summer.

If you approach Ixworth on the A1088 (O.S. map 211), just before you reach there across the Black Bourne is the watery area known as Mickle Mere. Much of the year it is flooded and is a good place to view wintering geese and other wildfowl. A track known as Baileypool Lane runs up the side of the Black Bourne and behind Pakenham Watermill.

Heveningham Park (O.S. map 231) in front of the magnificent Hall near Halesworth has recently had its lakes dredged, landscaped and restored to their former glory. A right of way runs right across the front of the grounds and crosses the long lake. This just has to be seen.

Not far from Southwold there are two splendid watery areas to investigate. Hen Reedbed (O.S. map 231) is a marshland nature reserve, created in 1999. You can see marsh harriers, bearded tits and a wide variety of other birds and insects. There is also a large heronry. There is a small amount of parking, and the track takes you from there over the road and along the levee beside the River Blythe. You can go most of the way to Southwold along Wolsey's Creek and Reydon marshes.

Close by, and to the south of the Henham Park Estate, you can seek out the flooded pit at Side Hill. This is a strangely beautiful place. leaving the A12 just north of the Soutwold turn, you can find a small track. This becomes a rougher track as it takes you past Park Farm and

plunges downhill to what looks like a perfectly natural feature, but is really a flooded pit. This is a largely unknown beauty spot and well worth a visit as these pictures show.

Lopham and Redgrave Fen (O.S. map 230) straddle the Norfolk-Suffolk border in the wet marshy area that is the source of two rivers. The Waveney rises here and flows east. The Little Ouse flows the opposite way. The Centre, run by the Suffolk Wildlife Trust provides parking, toilets and, at weekends, a teashop. There is a choice of walks around pools and reedbeds where you may be lucky enough to encounter the rare raft-spider, otters and water voles amongst a host of possibilities.

Coot

Trimley Marshes (O.S. map 197) near Felixstowe demonstrate that wildlife can thrive close to heavy industrialisation. The lagoons and islands are nesting sites for avocet, ringed plover and tufted duck. Winter wading birds flock here in large numbers. There is parking about a mile from the reserve, but walking is easy and a visitor centre opens at weekends.

Alternatively, you might park beside Shotley church and walk down past the house quaintly named 'number one below the church' before crossing water meadows to come to marshland right opposite Trimley Marshes. It is likely that here you'll encounter just as many birds and rather fewer bird-watchers.

Sedge Warbler

Slightly more out of the way, Hopton Fen and Market Weston Fen near Stanton (O.S. map 230) attract sedge and reed warblers. Nightingales nest here in summer. Lizards may be seen basking in the sunshine and marshy plants abound. Both of these wetland areas have paths that takc you close to the watery edges.

At Playford near Ipswich, if you walk along the footpath behind Hill House Farm, you may be able to enjoy limited access to the marshes around Playford Mere, home to snipe and other marshland wildlife.

Some of our Suffolk golf courses use water, both as a way of enhancing the area and as a hazard to the less fortunate golfer. There are several of these I might mention, but Ufford Park is particularly lovely, and home to a wide variety of waterlife.

Stowmarket golf course is set in the valley of the River Rat and makes the most of the natural landscape. Well signposted footpaths run across the park and into the damp woodland beyond. There is every chance of seeing kingfishers here and finding weird and wonderful fungi growing on the fallen tree stumps.

In Conclusion

Suffolk has host of damp corners. Plenty of them in this book will be familiar enough. But some of them come at you by surprise. They may be tucked away down roads that don't seem to go anywhere in particular, or not down roads at all.

I always worry that by revealing secret corners in a book like this, they won't remain very secret any more. But then, a lot of these spots only exist because enough people cared enough about the watery places in our bit of the world to protect and improve them.

So, it would be a bit mean not to tell you about these places, wouldn't it?

Sotterley church is buried in the heart of the Sotterley Hall estate. It is reachable by a path described on **www.suffolk churches.co.uk.**, or just by following the signs on the trees. The joy of this walk is that you get to enjoy the magnificence of the lakeland landscaping that is part of the setting. It's about a mile in each direction and involves stiles, gates and bridges, but it's well worth the effort. And almost as an added bonus, you'll find St. Margaret's church is one of the loveliest in Suffolk.

Sotterley

About halfway between Sudbury and Long Melford is the tiny wetland nature reserve known as Rodbridge (O.S. map 196). On the Suffolk side of the Stour, it has parking and a picnic area. The valley walk from there back to Ballingdon Bridge follows an old disused railway track and offers good views of the valley either side.

Following the footpath from Newbourne or Hemley down to the River Deben, you can continue downriver to Kirton Creek. Wet and wooded, this is one of my favourite places in the whole world. Not so long ago, I witnessed a nesting oyster-catcher driving off a marsh-harrier close to this marshy pool.

There are places already mentioned that I might not have emphasised strongly enough - Wissington (see page 11), Livermere (see page 22) and Benacre (see page 32), fortunately are relatively unknown and unspoilt.

But there are more. I have lived in Suffolk for most of my life and I'm still discovering those special places.

If you make your way to Brandon, (O.S. map 229) just across the bridge (on the Norfolk side) is a river path to the west that follows the Little Ouse as far as Hockwold. Here, you soon lose all hint of the twenty-first century (except for when you cross the railway). You can walk for three miles and not meet another human being.

Brandon

An equally lovely and equally deserted spot is Denham Castle. Little remains of the old motte and bailey but for the moat, which in dry periods can be a bit lacking in water. Along a by-way between Higham and Barrow (O.S. map 210), is a path that will lead you to this spot; an ancient and rarely-visited site. But it has a wonderful air of mystery and antiquity about it.

At Chevington, only a few miles to the east is one of the most impressive moats in Suffolk. Chevington Hall does not welcome visitors, but from the churchyard beside it, the moat around the site of the medieval hall (long since gone) can be seen in all its splendour. It is best viewed in winter, when the trees are bare. The signs at the edge of the churchyard warn of 'Deep Water.' Too true: the moat plunges down a good number of feet below the church that stands beside it. It is hard to contemplate the effort that must have been required to construct it.

Chevington hall moat - no picture I was able to take could do it justice

Between Great Livermere and Timworth, you may discover a marked footpath that crosses a field and enters a patch of woodland. Though you can't see it at that point, you are close to Ampton Water. Over a century ago, people from Bury came here in the winter to skate on the frozen lake. Now it is hard to imagine it ever freezing hard enough to be safe to skate on. If you continue through the woodland for about half a mile, you will reach a footbridge across the water, giving you wonderful views of the water birds that flock there.

At one edge the decaying tree stumps look rather like the everglades. You can then continue uphill and down again to the Broad Water at Livermere (see page 22). It is a beautiful place at any time of year and there is every likelihood you will have it all to yourself.

In short, the watery corners of Suffolk are a delight to anyone who takes the trouble to seek them out...

Why not explore Outney Common, just north-west of Bungay, preferably by canoe. Or take the footpath from Playford that leads you across the railway and to a hill overlooking Playford Mere. Or wander down Mill Lane at Great Blakenham past the old lock to the River Gipping. Or visit St. Peter's Brewery, set in a magnificent moated manor house. The choices are endless.

You can enjoy the Suffolk coast in all weathers. From Landguard Fort to the Ferry two miles to the north, Felixstowe at any season has a lot to offer. Better still, see the Suffolk coast from the sea.

The Balmoral leaving Southwold Pier

So why not seek out these wonderful watery corners.
...From the mysterious empty beauty of Benacre Broad with its terns and avocets....

©Richard Weale Wildlife Photography

...to the crumbling cliffs of our most deserted pieces of coastline...

..to the lovely Lark Valley between Culford and Mildenhall...

©Richard Weale Wildlife Photography

And you might just be privileged enough to experience a moment of magic such as this egret caught by the camera of Richard Weale at the Dingle marshes near Dunwich.

These are a few of my favourite places. But you may know of others I've neglected to include. Assuming this book may be revised and reprinted, any suggestions will be welcomed. Please send your ideas to: **pipandjulie14@talktalk.net.**